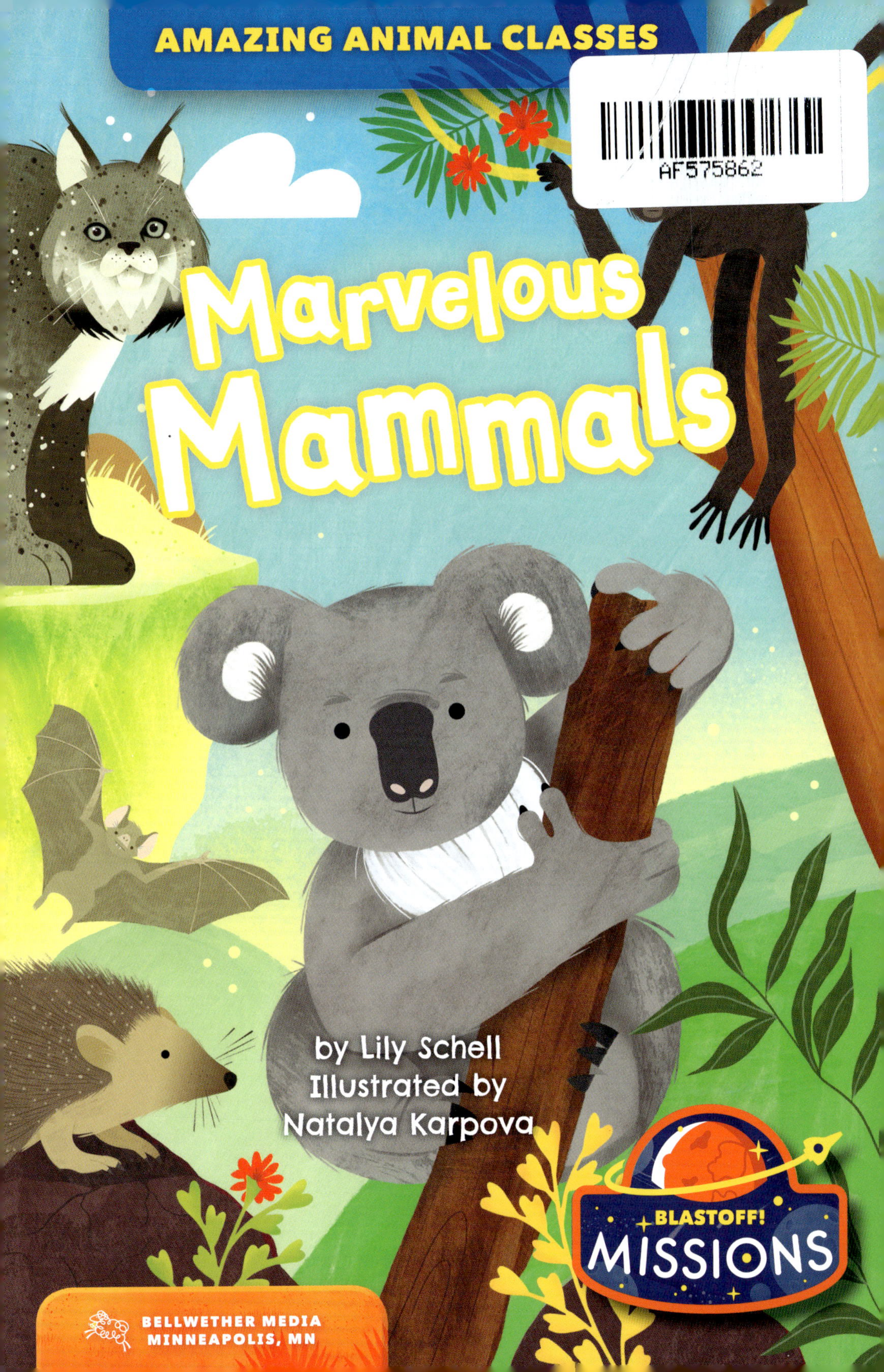
AMAZING ANIMAL CLASSES
AF575862
Marvelous Mammals
by Lily Schell
Illustrated by
Natalya Karpova
BLASTOFF!
MISSIONS
BELLWETHER MEDIA
MINNEAPOLIS, MN

Blastoff! Missions takes you on a learning adventure! Colorful illustrations and exciting narratives highlight cool facts about our world and beyond. Read the mission goals and follow the narrative to gain knowledge, build reading skills, and have fun!

Traditional Nonfiction

Narrative Nonfiction

Blastoff! Universe

MISSION GOALS

- FIND YOUR SIGHT WORDS IN THE BOOK.
- IDENTIFY THE TRAITS OF MAMMALS.
- THINK OF QUESTIONS TO ASK WHILE YOU READ.

This edition first published in 2023 by Bellwether Media, Inc.

Library of Congress Cataloging-in-Publication Data

Names: Schell, Lily, author.
Title: Marvelous mammals / by Lily Schell.
Description: Minneapolis, MN : Bellwether Media, Inc., 2023. | Series: Blastoff! Missions. Amazing Animal Classes | Includes bibliographical references and index. | Audience: Ages 5-8 | Audience: Grades 2-3 | Summary: "Vibrant illustrations accompany information about mammals. The narrative nonfiction text is intended for students in kindergarten through third grade"-- Provided by publisher.
Identifiers: LCCN 2022020222 (print) | LCCN 2022020223 (ebook) | ISBN 9781644876510 (library binding) | ISBN 9781648348358 (paperback) | ISBN 9781648346972 (ebook)
Subjects: LCSH: Mammals--Juvenile literature.
Classification: LCC QL706.2 .S333 2023 (print) | LCC QL706.2 (ebook) | DDC 599--dc23/eng/20220429
LC record available at https://lccn.loc.gov/2022020222
LC ebook record available at https://lccn.loc.gov/2022020223

Editor: Betsy Rathburn   Designer: Andrea Schneider

Printed in the United States of America, North Mankato, MN.

Table of Contents

On a Mammal Mission!

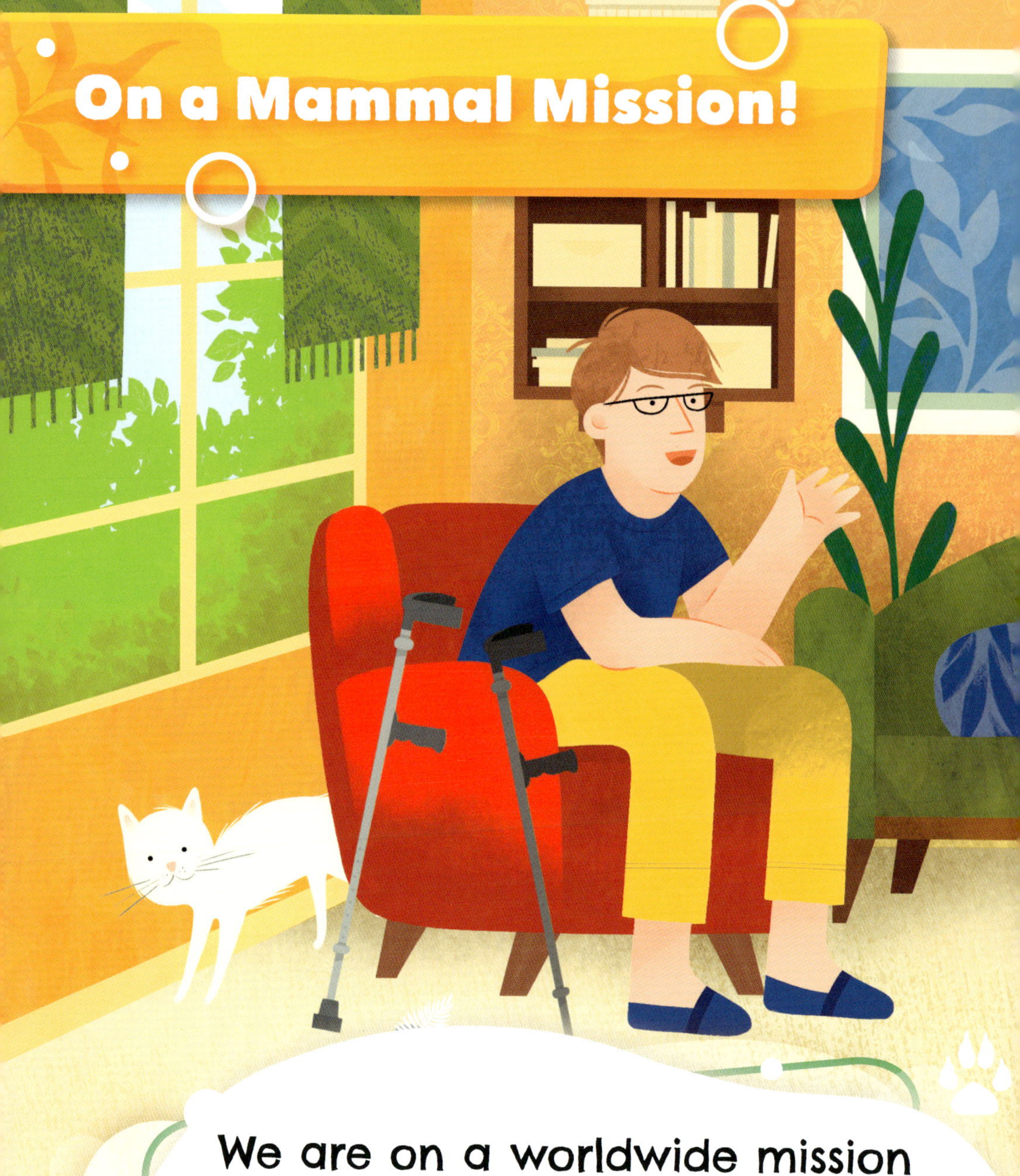

We are on a worldwide mission to find **warm-blooded** animals with backbones and hair or fur. Their babies drink milk. We are looking for mammals!

Look around your home.
Pet dogs and cats are mammals!

Snow Sprinters

Welcome to Canada's **taiga**! Do you have **snowshoes** on?

These mammals do not need them. That lynx chases snowshoe hares in deep snow. These animals have big feet to keep from sinking!
snowshoe hares

In the Forest

South America's Amazon **Rain Forest** has many mammals. Look up! Spider monkeys swing in the trees.

We should leave before it gets too dark. Vampire bats are starting their hunt!
vampire bats
JIMMY SAYS
Bats are the only mammals that can fly.

Across the ocean, Australia's open forests host many **marsupials**. This koala hugs tree branches here.
koalas

Below, a mother kangaroo hops past. She carries her baby in her pouch!
kangaroos
pouch
JIMMY SAYS
Australia has many unusual mammals. Platypuses and echidnas are monotremes. These mammals lay eggs!

Keep your eyes peeled as we head to Asia's Himalayan Mountains. A master of **camouflage** lives in these forests.

This furry red panda matches
the moss in its treetop home!

Prowling Predators

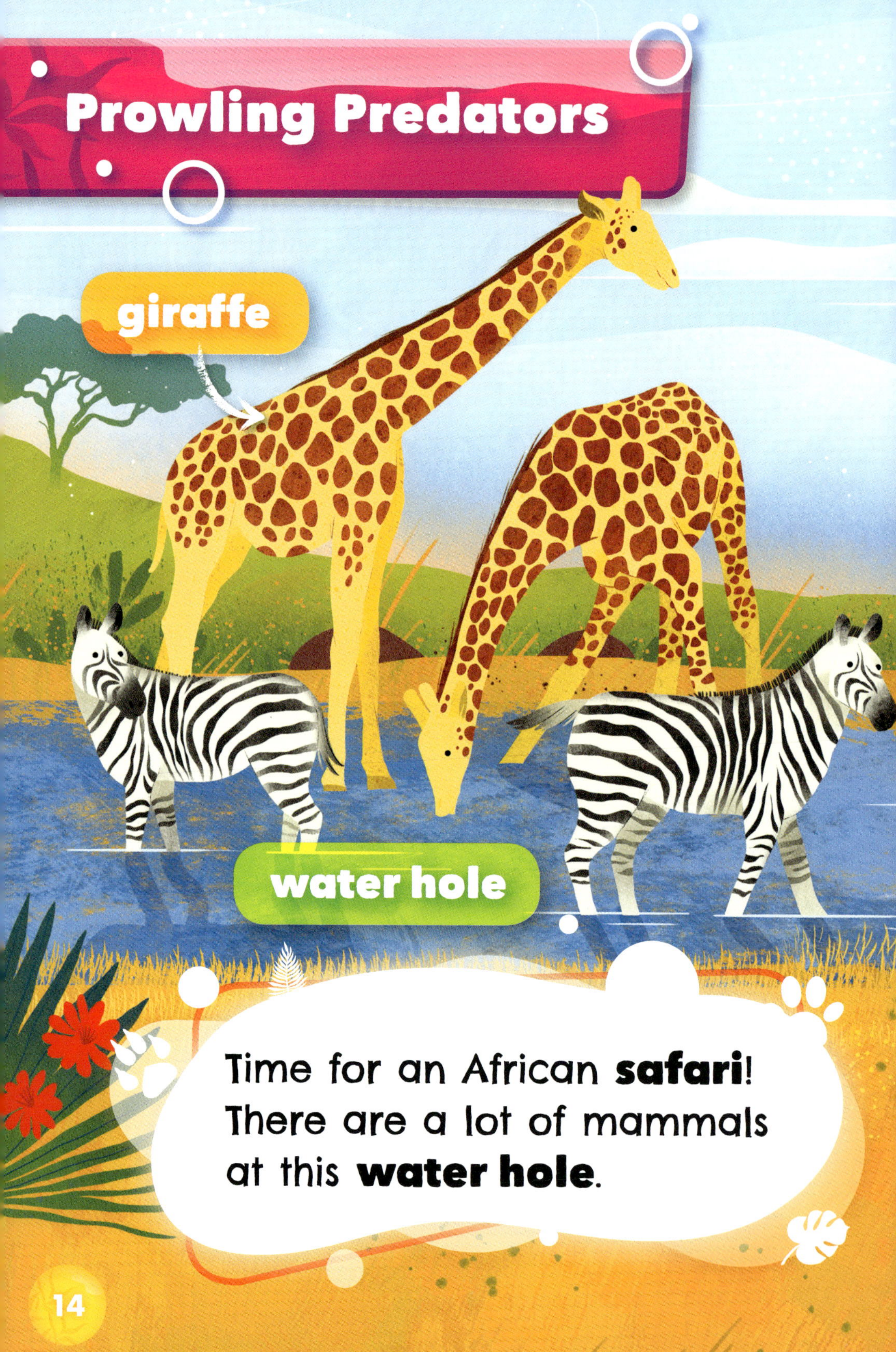

Time for an African **safari**! There are a lot of mammals at this **water hole**.

The giraffes bend down far for a drink. That **herd** of zebras watches for **predators**. Uh-oh. Here come hyenas!

Let's head to the English countryside. See the fox prowling for food? That red squirrel above chatters a warning. The hedgehog runs to avoid being dinner!
fox

red squirrel
hedgehog

Underwater Wildlife

pod of dolphins

Dive underwater to meet some ocean mammals! Here comes a **pod** of playful dolphins.

Wow! Look at that blue whale.
It is the largest animal on Earth!

We nearly forgot one mammal on our mission. It is you! Yes, humans are mammals, too! What other marvelous mammals can you name?

Mammal Facts

have backbones

have hair or fur

feed babies milk

Glossary

camouflage—a way of using color to blend in with surroundings

herd—a group of animals that live and travel together

marsupials—mammals that carry their young in a pouch on their bellies

pod—a group of dolphins that live and travel together

predators—animals that hunt other animals for food

rain forest—a thick, green forest that receives a lot of rain

safari—a trip to see animals in the wild

snowshoes—large, wide frames that attach to shoes to keep feet from sinking into snow

taiga—a northern forest that has mostly evergreen trees such as spruce and fir

warm-blooded—having a high body temperature that does not change with the environment

water hole—an area of water where animals gather to drink

To Learn More

AT THE LIBRARY

Daniels, Patricia. *Mammals.* Washington, D.C.: National Geographic, 2019.

Kenah, Katharine. *Super Marsupials: Kangaroos, Koalas, Wombats, and More.* New York, N.Y.: HarperCollins Publishers, 2019.

Labrecque, Ellen. *Do Vampire Bats Really Drink Blood? Answering Kids' Questions.* North Mankato, Minn.: Pebble, 2021.

ON THE WEB

FACTSURFER

Factsurfer.com gives you a safe, fun way to find more information.

1. Go to www.factsurfer.com.

2. Enter "marvelous mammals" into the search box and click 🔍.

3. Select your book cover to see a list of related content.

BEYOND THE MISSION

> CREATE A MAMMAL. WHAT DOES IT LOOK LIKE? DRAW A PICTURE.

> WHICH MAMMAL WOULD YOU LIKE TO LEARN MORE ABOUT? WHY?

> CAN YOU THINK OF ANOTHER MAMMAL THAT WAS NOT MENTIONED IN THE BOOK?

Index